THE INCREDIBLE WORLD OF PLANTS

PLANTS UNDER THE SEA

CHELSEA HOUSE PUBLISHERS
New York • Philadelphia

Text: Andreu Llamas
Illustrations: Luis Rizo

Plantas del mar © Copyright EDICIONES ESTE, S. A., 1995,
Barcelona, Spain.

Plants Under the Sea copyright © 1996 by Chelsea House
Publishers, a division of Main Line Book Co. All rights
reserved.

1 3 5 7 9 8 6 4 2

Library of Congress Cataloging-in-Publication Data

Llamas, Andreu.
  [Plantas del mar. English]
  Plants under the sea / text, Andreu Llamas ; illustrations,
Luis Rizo.
    p. cm. — (The Incredible world of plants)
  Includes index.
  Summary: Describes plants that live under the sea and dis-
cusses their ability to adapt to unique situations such as scarcity
of light and underwater caves.
  ISBN 0-7910-3468-2. — ISBN 0-7910-3474-7 (pbk.)
  1. Marine ecology—Juvenile literature. 2. Marine plants—
Juvenile literature. [1. Marine ecology. 2. Ecology. 3. Marine
plants.] I. Rizo, Luis, ill. II. Title. III. Series: Llamas, Andreu.
Incredible world of plants.
QH541.5.S3L5813   1996   95-18261
574.5'2636—dc20   CIP
                                                   AC

CONTENTS

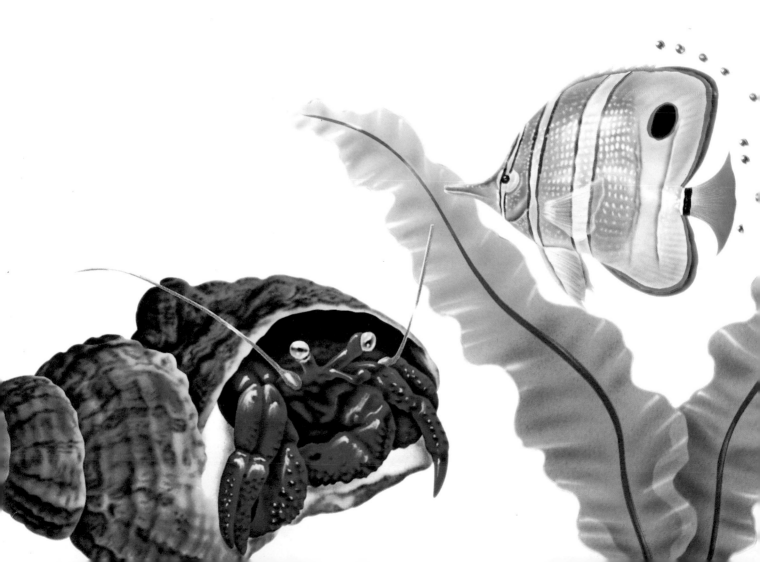

THE AREAS OF THE COAST

The place where the land meets the sea is called the *seaboard*. Here the plant and animal life is different from those in the sea, and these life forms cannot live on land or in fresh water.

Two-thirds of our planet is covered with water, so the length of the coasts is enormous (although the shore may only be a few feet wide).

For millions of years and thousands of times every day, the waves raised by the wind break incessantly against the shore. During storms the waves can be very strong. This is why only a special group of plants and animals have adapted to living there; for instance, the plants that live near the shore have to be able to withstand strong winds and salinity, so they are usually quite short.

The coast may have rocky cliffs or form very flat sandy beaches; however, all coasts are divided into strips that are more or less parallel to the sea. Each one has different animals and plants from those that live on higher or lower strips.

The strips include the splash zone, the upper shore, the middle shore, and the lower shore. The highest is the splash zone, where only a few drops of water reach from time to time carried by the wind. The upper shore is only covered by water at high tide, while most of the middle shore is covered and uncovered each day. The lower shore is only uncovered at low tide. The area below this, the sublittoral, is under the seaboard and is never uncovered.

(1) Distribution of terrain
Each strip of the coast has its own characteristic inhabitants.

(2) Protection from the wind
To protect itself from the wind, the sea carnation grows in tight, compact formations.

(3) Fixing the sand
Some plants have dragging stalks and strong roots that help to stabilize the sandy soil.

(4) Hunter worms
In the rocky pools of the lower shore many sea worms can be found. The best-known are the nereis, which are terrible 8-inch (20-centimeter) long hunters with strong jaws.

(5) Digging a shelter
The common sea urchin (Paracentrotus lividus) can dig into a rock by moving its strong spikes back and forth and by using its teeth. In the end it manages to make quite a large hole.

(6) Sea dates
Sea dates live on rocks and dig passageways. To do this they secrete a chemical substance that dissolves the rock.

4

LIVING WITH THE TIDE

The tides are the rise and fall in the sea level due to the attraction of the moon and sun. The tide rises twice every 24 hours.

Life on the coast is marked by sudden changes in atmospheric conditions of humidity, temperature, and light. The force of the waves also requires creatures to latch on hard to whatever they can grasp. Some creatures, such as barnacles, stick to rocks as adults and develop conic shells to protect themselves.

Rocky coasts battered by the waves are cracked and rough and contain a huge number of seaweed, such as coralline and sea lettuce, and animals such as the sea tomato, mussels, limpets, periwinkles, and crabs. Some of these animals, like the mollusks, are able to drill their own tunnels and cavities in the rocks to escape from the strength of the waves. Most of them also have very hard shells to protect themselves from the waves, from predators, and from the heat of the sun that dries them. Many animals search for food in these rich surroundings, such as seagulls to sargos and giltheads, which are capable of breaking the limpets' shells.

Plants and animals on the upper shore are out of the water most of the time. By contrast, those on the lower shore are only uncovered for a short while before the tide begins to rise again.

Mussels
Mussels hold on tight to the rock with special filaments and when the tide comes in they open their shells to obtain food by filtering the sea water.

Limpets
Limpets hold on to the rock with their muscular foot. Each day when the tide comes in, limpets can move 8 inches to look for food but return to their positions when the water goes down again. They hold on so tight that they leave marks on the rock.

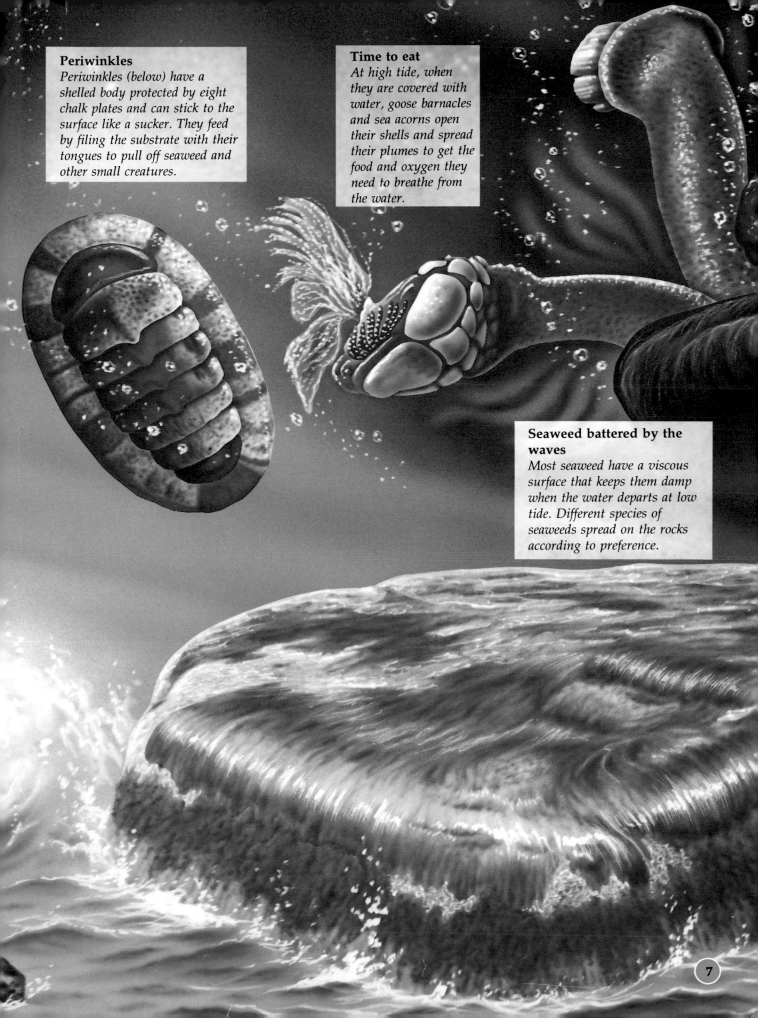

Periwinkles

Periwinkles (below) have a shelled body protected by eight chalk plates and can stick to the surface like a sucker. They feed by filing the substrate with their tongues to pull off seaweed and other small creatures.

Time to eat

At high tide, when they are covered with water, goose barnacles and sea acorns open their shells and spread their plumes to get the food and oxygen they need to breathe from the water.

Seaweed battered by the waves

Most seaweed have a viscous surface that keeps them damp when the water departs at low tide. Different species of seaweeds spread on the rocks according to preference.

SEAWEED

Seaweed are plants that have adapted to life in the sea. Just like other plants, seaweed use sunlight to produce food, so they only survive in well-lit areas. However, they are very different from other plants—they have no roots, stalks, leaves, flowers, or fruit.

Instead of roots, seaweed have bases or fingerlike disks that enable them to stick to the roughness of rocks, but which cannot pick up nutritious substances. This is not a problem, however, as the sea gives seaweed all the minerals they need to grow, which are absorbed directly through the surface of the plant. They do not need flowers or fruit, because new seaweed are born from small male and female parts that are released in the water. The largest seaweed have stems and foliage.

Seaweed are not stiff enough to stand up out of the water, but that does not mean they are not strong, as their shape is designed to stand the strong waves.

If you touch them, you will find that seaweed are slippery because they are covered in a jellylike substance that prevents them from drying out under the sun and wind.

The seaweed that live closest to the surface are usually green, like land plants, whereas at greater depths brown and red seaweed abound. The appearance of seaweed at these depths changes with the seasons, since seaweed and animals develop most in spring and summer.

Seaweed produce repellent substances to protect themselves from predators, but are still devoured by sea urchins, sea hares (aplysia), loaches, and a large amount of worms, snails, and crustacea which in turn attract small fish such as sea slugs, gudgeons, and doncellas.

(1) Animals or plants?
On the seabed, animals and plants can disguise their forms. Ancient scientists mistakenly classified many animal species as plants.

(2) Fish or seaweed?
There are fish like this sea dragon in tropical seas that have changed their bodies to imitate pieces of floating seaweed. This is how they deceive predators.

(3) Sea lettuce
Sea lettuce are seaweed of 6 to 12 inches (15 to 30 centimeters) long that live on rocks in the area between tides, especially in waters rich in organic material.

(4) Starfish
Starfish have hundreds of small mobile feet under their arms, which transport them and enable them to hold on to the rock.

(5) Nudibranchiate eggs
The masses of nudibranchiate eggs (mollusks without shells) look like striking tapes. But they are protected—the eggs contain a substance with a very unpleasant taste.

(6) Acetabularia
Here you can see how acetabularia seaweed look. They can grow to 3 inches (8 centimeters) long but the seaweed is a single cell!

"FLOWER ANIMALS"

Sea anemones are animals that live on almost all rocky coastlines and even in docks and on breakwaters. Their bodies are soft and flowerlike but they are really predators that feed by catching prey in their many tentacles.

Sea anemones, just like jellyfish, belong to the Cnidarie family, which are characterized by having a curious defense and attack mechanism, the cnidoblasts. These are specialized cells that act as poisonous prickles and are used in defense to paralyze the victim. When a fish comes too close to a sea anemone and touches one of its tentacles, a large number of poisonous filaments are shot out and rapidly paralyze the fish. The sea anemone can then take it to its mouth and eat it in peace.

When the tide falls, most anemones that live near the coast bring in their tentacles and become jellylike balls to avoid drying. They stay like this until the sea covers them again. The prickles of anemones that live in cold water seas are not very strong (they cannot penetrate human skin), but the ones that live in the warm tropical seas are far more irritating. There are giant anemones in tropical seas up to 3 feet (1 meter) in diameter.

A house with tentacles
Clown fish have learned to live among the tentacles of anemones, moving calmly without being eaten and without being affected by the poison. There, their home is well protected by the tentacles.

Sea feather
Despite its name, the sea feather is also an animal colony made up of tiny bodies.

It is not a flower
The spirograph is really a beautiful worm that lives inside a tube it makes. It spreads its filaments to get food and oxygen but it hides them at the slightest sign of danger.

Anemones
Anemones have up to 170 tentacles to catch prey after paralyzing them with the poison from their stinging cells.

Sea tomato
Here you can compare the appearance of an actinia, or sea tomato, both in and out of the water. When the water level falls, the actinia fold their tentacles to avoid drying.

Gorgonians
Gorgonians are animals that form fanlike colonies that also look like plants.

11

SEAWEED AND LIGHT

At depths greater than 33 to 50 feet (10 to 15 meters), the underwater scenery begins to notice the scarcity of light.

Seaweed colonizing these environments adapt to the darkness and cover the bed in strips in which a few species of green and red seaweed predominate.

On the seabed there may be slanted rocks and great blocks of stone that have broken off due to erosion and whose lower faces house a rich fauna of colonial animals scattered among the seaweed, especially sponges and ascidians, as well as worms, snails, and mollusks.

Cavities give life to crustacea such as lobsters and territorial fishes, from small fish to great hunters such as conger eels and groupers.

Here the scenery changes little throughout the year but the small creatures that live on the seaweed do have a seasonal cycle; this is why the seaweed are loaded in spring and look hairy, in autumn they become whitish, and in winter they are green and clean.

Seaweed can only live where they can get enough light; no plants survive beyond 328 and 656 feet and only animals live at great depths. Not all seaweed are the same color, since it depends on their colored pigments, as well as *chlorophyll*. Green seaweed grow near the surface, whereas brown and red seaweed live in deeper waters.

Hard seaweed
The cellular walls of some seaweed are calcified so instead of being flexible, they form hard pieces that do not resemble our image of seaweed.

Ball-shaped seaweed
Codium bursa is a strange seaweed with filaments that intertwine to form a soft spherical thallus.

Fishing with a rod
The angler fish is an amazing hunter; one of its prickly radii ends in a fleshy portion that it uses as bait to attract small fish to its mouth. To deceive its victims it moves the filament as though it were a worm on the end of a fishing line.

The fearful Portuguese caravel
The Portuguese caravel, or man-of-war, is one of the most dangerous jellyfish, because they have very long tentacles (up to 49 feet long) full of stinging cells that can cause total paralysis and almost instant death in any prey.

Good partners
Some hermit crabs transport one or more anemones on their shells. The crab uses the camouflage and protection of the poisonous tentacles while the anemone gains a comfortable means of transport.

A good hiding place
This seaweed at right, almost completely hardened, form many holes that serve as a refuge for many animals of the sea.

UNDERWATER CAVES

In underwater caves, the darkness prevents vegetable life from developing. Further into the cave, there is less fauna on the walls, and the floor is practically bare. However, in tunnels, due to the water movement, the walls are usually covered with abundant fauna and filtering animals.

However, underwater caves are not devoid of life. Living on the floor are holothurians, ophiurans, and crabs, and between the rocks and stones live many species of crustacea, such as lobsters.

Moreover, not all the animals live alone or in small groups. One of the most spectacular of these animals is the tiny misidaceous crustacea, which form a thick cloud in the corner of caves and are food to many other crustacea and fish, such as kinglets.

Caves also serve as hideouts for hunters that prefer to stalk at night, such as conger eels. Very rare species, which are actually deep water creatures, also live in these caves. Although there are no seasonal changes in the caves, there is great contrast between day and night. During the night almost all of the cave inhabitants leave their territory to feed outside under the cover of dark.

(1) A mysterious world
At the entrance of underwater caves, the light permits the development of vegetation, but the deep and mysterious interior is practically void of all types of plant life.

(2) Living facedown
These crustacea live on the rocky littoral between cracks and stones from 16 to 98 feet in depth. They are often seen facedown, stuck to cave roofs.

(3) The kinglet
Kinglets, or apogon, almost always live hidden by forming groups in small holes or caves. They feed during the night, leaving their refuge to hunt small fish and invertebrates.

(4) Filtering the bed
Holothurians feed by taking in sand and sediments from which they extract food and throw out the rest leaving piles.

(5) The lobster
They hide in cracks and holes in the rocks between 66 and 230 feet underwater and only stick out part of their head and antennae. When winter months come, lobsters migrate to greater depths.

(6) Ascidians
Despite their appearance, ascidians are really animals. They feed by filtering the water they breathe in through the syphon.

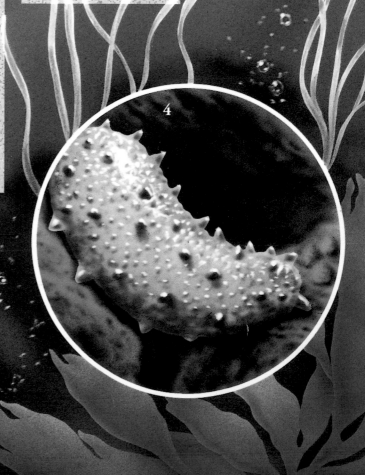

4

FORESTS OF WRACK

Seaweed communities reach their greatest splendor on the hard coasts of cold and temperate seas, where they form impressive "forests" of wrack.

Wrack is enormous seaweed with a base disk strongly fixed to the substrate. Instead of a stalk they have a *stipe* with one or more filoids (leaves). This seaweed may be over 130 feet (40 meters) long.

The wracks that make up the most spectacular underwater forests are the laminaria and the macrocystis. The macrocystis communities are the largest. Base disks support stipes of up to 164 feet (50 meters). Wrack forests are far less impressive—their stipes only reach 3 to 13 feet (1 to 4 meters) in length.

The filoids of this seaweed have air bubbles that enable them to float on the water and form a thick layer of vegetation on the surface. The seaweed grows very fast, from 1 inch (3 centimeters) per day in laminaria to 20 inches (50 centimeters) per day in ideal conditions for macrocystis.

Many animals live in wrack forests. Sea otters swim on the surface and submerge to look for food. When they want to sleep they wrap themselves in some seaweed on the surface to avoid being swept away by the current when they are sleeping. However, wracks do have enemies. One of the most active predators of wracks is the sea urchin.

Wrack forests show seasonal changes, but not drastic changes like garweed meadows, because the seaweed never lose all their leaves.

(1) Seaweed larger than trees
Macrocystis are giant garweeds from California that grow spectacularly. Some grow up to 3 feet (1 meter) in one day and can reach 328 feet (100 meters) in length.

(2) Tunnel-shaped bodies
The salpas' bodies look like a transparent tunnel that measures up to 4 inches (10 centimeters) long. They sail on the open sea and can form long chains.

(3) Crinoids
Crinoids have five pairs of ramified arms, and they are the same type of animals as starfish.

(4) Air bubbles
Thanks to air bubbles in its interior, the laminaria can stay straight up under the water.

(5) Hydrozoan colonies
Hydrozoans are tiny animals that form colonies a few inches high; many live on seaweed.

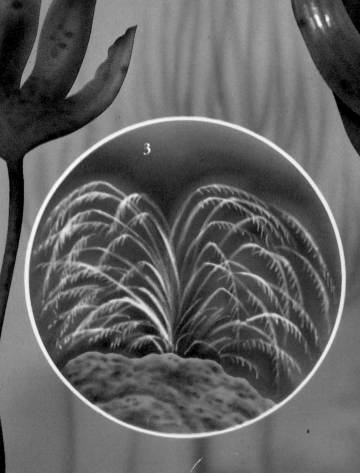

GARWEED MEADOWS

Garweed is phanerogamous, that is, a plant with roots and leaves. It forms underwater meadows made up of blankets of long tapelike leaves less than half an inch wide and up to 3 feet long. Many animals hide in the grass and many small seaweed and animals become encrusted on the leaves.

Garweeds have a lot of cellulose so very few species eat them, except sea urchins and fish. The sediments that accumulate on these meadows are much richer in organic material and mineral salts than sandy floors, and many animals that live in suspension take advantage of the nutrients, as well as sedimentivores, which are also the food of carnivorous animals. Garweeds have an advantage over seaweed: they have underground stems called rhizomes that give the substrate stability and allow the meadow to extend both vertically and along the ground, and they can act as reserve organs. They also have roots to get down to the bed where there is a greater concentration of nutrients. Their leaves are specialized in production and they can provide for the rest of the plant thanks to their complete circulatory system.

Garweeds have a very clear annual cycle. Each autumn leaves fall off, but in the winter new leaves begin to sprout from the rhizomes. During the spring, leaves grow green and strong, and as they develop they become encrusted with small creatures. This is why at the end of summer the leaves look white and bend under the weight of the encrustations.

(1) Underwater meadows
Garweed meadows cover enormous extensions if conditions are right.

(2) Sea balls
Broken garweed leaves dragged along by the waves, mix with grains of sand and form soft balls which once were believed to be the fruit of the plant.

(3) Nudibranchia
They look like underwater flowers, but they are really animals capable of feeding themselves with food that others may not approach. For example, stinging nematocysts do not bother them.

(4) To eat and be eaten
Sea urchins are one of the few animals able to eat garweed, but they also have their enemies. Here you can see a triton calmly eating a sea urchin while it is protected by its shell.

(5) Sea horses
Sea horses are very curious fish. They swim by vibrating their pectoral fin up to 70 times per second, and the males have the responsibility of transporting and looking after the eggs in a bag on their bodies until the babies are born.

(6) A good support
Garweed leaves maintain many other creatures like these small epiphyte seaweed.

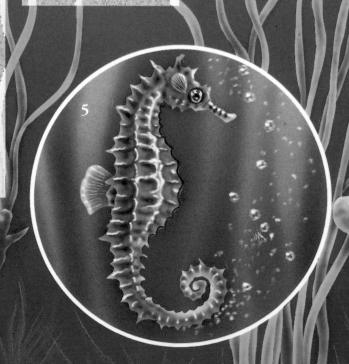

5

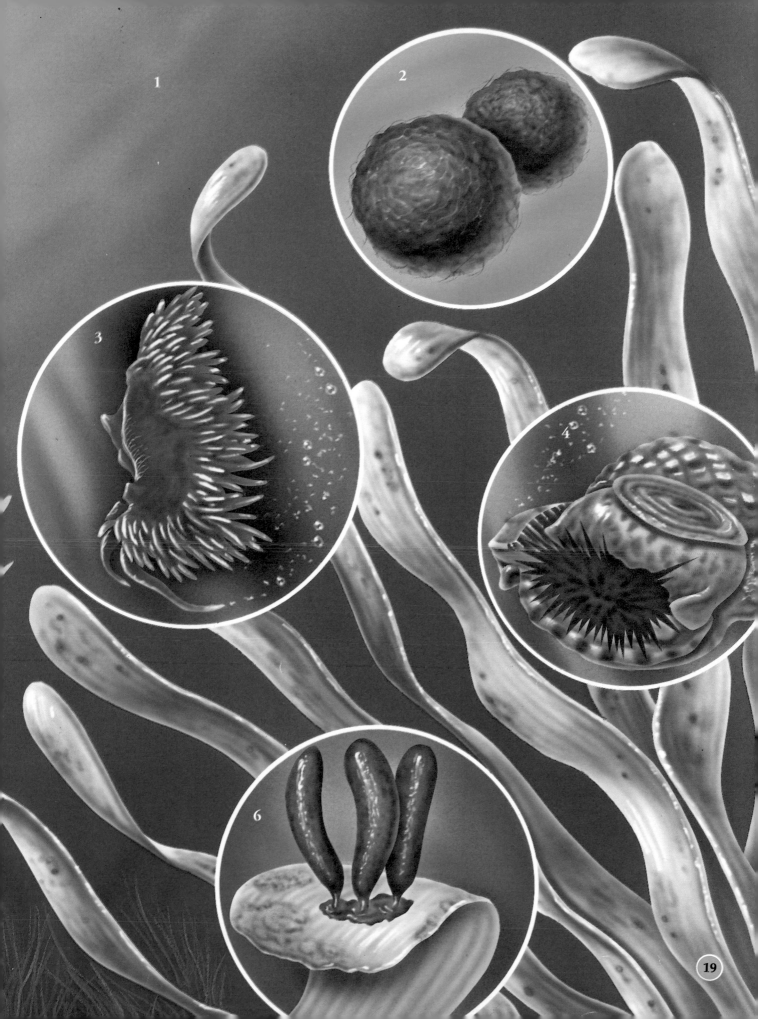

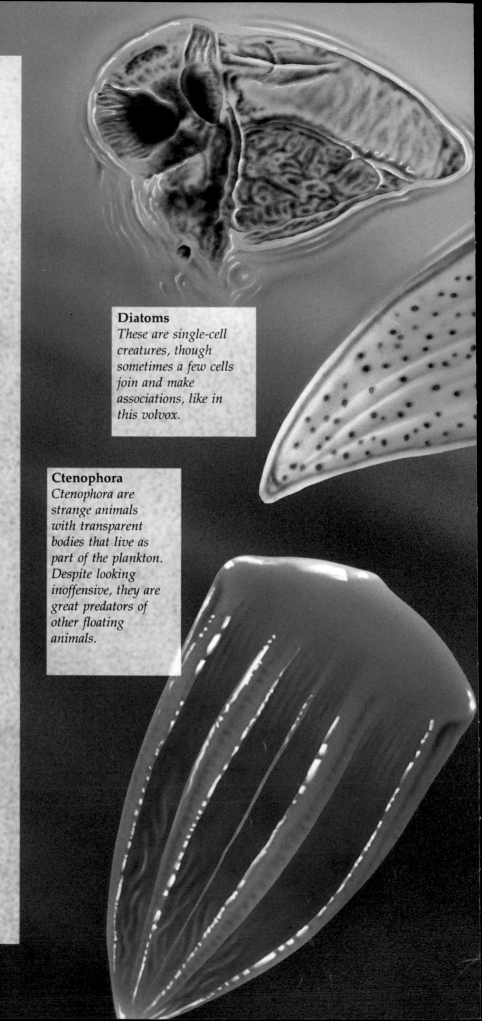

TRAVELING PLANKTON

There are billions of tiny plants and animals in the sea that do not swim but are rocked by the currents. They are plankton.

Most plankton are smaller than 1 millimeter, but are essential for life in the sea. They are the staple diet of many fish, seabirds, and whales.

There are two main type of plankton: phytoplankton (vegetable) and zooplankton (animal). Zooplankton includes a variety of animals—from tiny creatures to jellyfish several feet long. Vegetable plankton live near the surface because they need light for *photosynthesis*. They also prefer cold waters because they contain more minerals. Animal plankton feed on vegetable plankton.

The most abundant seaweed in the ocean are almost invisible; one quart of sea water may give life to millions of single-cell creatures that form part of the plankton. The most numerous of them are the diatoms, of which there are more than 10,000 different species. They also have an incredible reproductive capacity—one single diatom can produce 1,000 descendants per week.

We can also find fish eggs and young fish or other animals beginning life as part of the plankton, but then they grow and become much larger, such as crabs, starfish, and octopuses.

Diatoms
These are single-cell creatures, though sometimes a few cells join and make associations, like in this volvox.

Ctenophora
Ctenophora are strange animals with transparent bodies that live as part of the plankton. Despite looking inoffensive, they are great predators of other floating animals.

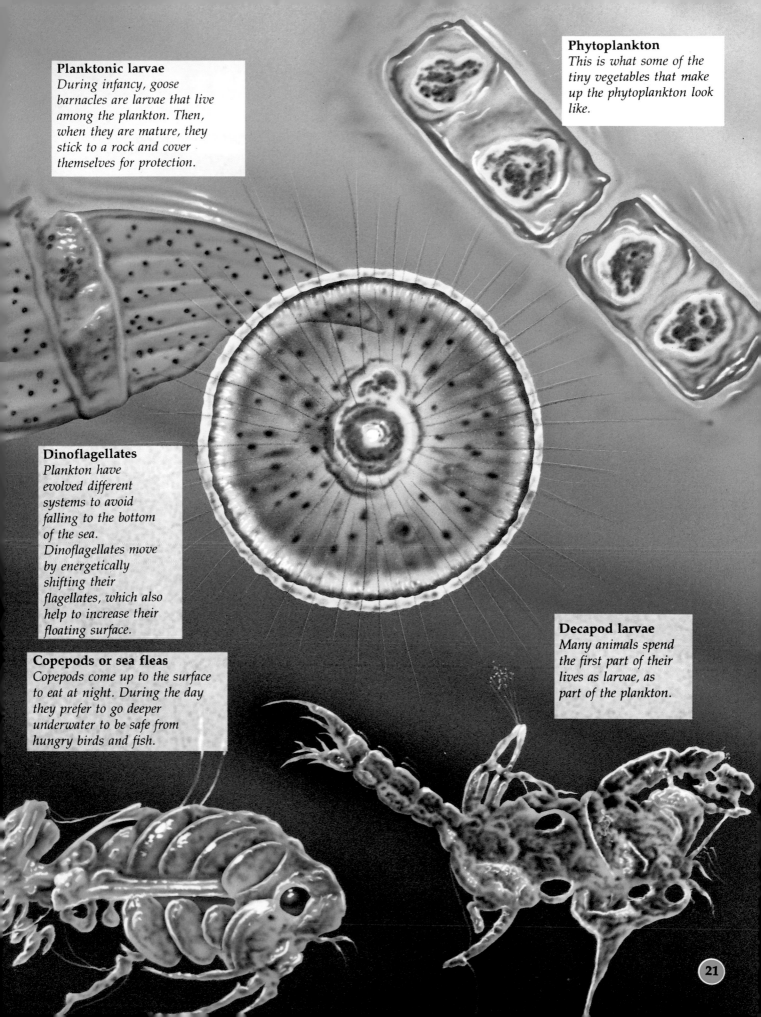

Planktonic larvae
During infancy, goose barnacles are larvae that live among the plankton. Then, when they are mature, they stick to a rock and cover themselves for protection.

Phytoplankton
This is what some of the tiny vegetables that make up the phytoplankton look like.

Dinoflagellates
Plankton have evolved different systems to avoid falling to the bottom of the sea. Dinoflagellates move by energetically shifting their flagellates, which also help to increase their floating surface.

Decapod larvae
Many animals spend the first part of their lives as larvae, as part of the plankton.

Copepods or sea fleas
Copepods come up to the surface to eat at night. During the day they prefer to go deeper underwater to be safe from hungry birds and fish.

THE MYSTERIOUS SARGASSO SEA

Sargassos are seaweed common along the coasts of Central America and the Caribbean Sea. When the terrible tropical storms begin, the force of the waves pulls them out of their supports and they float away out of control over the waves until they are caught by a strong current in the middle of the ocean, to the northwest of Bermuda.

The sargassos that float have created a new, very special ecosystem known as the Sargasso Sea. Here there are some incredible animals so well adapted to living only there that they receive such names as sargasso fish, sargasso crab, and sargasso sea horse.

However, one of the most spectacular inhabitants of this sea is the eel. Eels spend almost their entire life in the rivers of Europe and North America, but always return to the Sargasso Sea where they were born to *spawn* and die there. Their larvae are called leptocephalous and float among the sargassos until they are dragged away by the Gulf Stream on a three-year voyage to the continents.

Actually, the Sargasso Sea has very little population and very few species, because no current brings nutrients to the area. The phytoplankton are quite scarce, and there are very little zooplankton. Furthermore, sargassos are practically the only species of seaweed capable of reproduction in this mysterious sea.

The giant sargassos are the largest plants living under the sea. They can grow over 330 feet (100 meters), taller than a 15-story building!

(1) Underwater trees
Sargassos look like impressive underwater trees. The strangest creatures in the sea live around them.

(2) Noctilucas
These are single-cell bodies that can grow up to 1/8 inch in diameter. They live in temperate and warm seas and are able to produce light in the form of brief gleaming when the water moves.

(3) Floating
There are air pockets inside sargassos that enable the seaweed to float.

(4) Parasitic isopods
The isopods are a group of small crustacea that include wood lice. Many sea isopods live as parasites on the bodies of fish, sucking their blood while they are transported.

(5) Eels
Eels' lives are marked by their incredible migration from the rivers to the Sargasso Sea, where they were born years before, to lay their eggs. In the pictures you can see: (5a) eel spawn; (5b) larva; and (5c) an adult eel, living in the river before making the return journey to the Sargasso Sea.

4

1

2

3

5c

5b

5a

AN ISLAND GROWS

Coral reefs are incredible underwater gardens made up of colonies of millions of tentacled animals called coral.

Each member of the colony (polyp) lives inside a chalk case that it builds itself. When it dies it leaves its stone skeleton and another coral builds its home on top of the old one. After many years, these mounted structures form mounds called *reefs* that can even come out of the water and form islands.

Coral reefs are some of the largest structures created by living beings. Fossil reefs have been found that now make up mountain ranges on occasions very far from our present oceans, with reef layers up to 3,280 feet thick. Corals need clear, warm waters with few nutrients to be able to proliferate and form reefs. In order to live, these corals also maintain a *symbiotic* relationship with some single-cell seaweed called zooxanthellae.

A special kind of reef is the barrier reef, which is always far offshore and is in a line. The most well-known of these is the Australian Great Barrier Reef which is more than 1,200 miles (2,000 kilometers) long.

Corals do not adapt well to any climatic change, so hurricanes or tidal waves cause great damage. Corals have many predators, such as some snails and starfish.

There are small sponges that perforate the coral to make a complex system of tunnels that usually end with the life of the coral as they weaken its structure and make them brittle.

(1) Coral reefs
Coral reefs are one of the world's communities in which the most life exists. Thousands and thousands of individuals compete for space to survive, growing on top of each other until they reach the surface and form an island.

(2) The bryozoans
Bryozoans are animals that live by forming colonies in a beautiful network. They secrete chalk skeletons that contribute to the structure of the reef.

(3) Eating coral
The prickly starfish is a coral eater. From time to time their numbers increase and then they can cause great damage to the Australian Great Barrier Reef.

(4) A powerful poison
This cone is perilously armed. It has poisonous darts that it fires and sticks in its victims and enemies. It can even kill a man in a few hours!

(5) False eyes
Some fish have a false eye "painted" on their bodies to confuse predators that do not know where their prey's head is nor in what direction it will swim away.

(6) Bad taste
Some sea slugs have striking colors to warn possible predators of their bad taste. Furthermore, if an attacker pulls off one of its papillae, it gets a new one in a short time.

THE INCREDIBLE CORAL

For a long time, coral was considered to be plant life due to its branchlike structure. However, they are really animals that live together and form colonies.

Each member of the colony is a small sack with an opening surrounded by a tentacle that acts as a mouth. The tentacles catch small prey and the sack digests them like a stomach. Each polyp protects itself with a hard chalk cavity, which it has built itself, and coral forms as these polyps come together.

In a coral reef, the animals live on top of each other fighting for space, but not all species of coral form reefs. There are many species that grow as isolated colonies, as is the case of red coral of the Mediterranean.

At the beginning of their lives, the coral larvae join the zooplankton and are carried by the ocean currents. Coral species may travel over 620 miles.

The animals that live in coral reefs are very special. There are many fish that live as couples. They may last a reproductive cycle or until one of them dies, as is the case of many species of butterfly fish. Sex changes are also quite common: in a colony of clown fish where one female lives with three or four males, if the female dies the strongest male immediately becomes a female.

Solitary coral
Not all coral forms colonies or reefs. Solitary northern coral prefer to live in cold waters where they grow individually and very slowly.

Soft coral
Halcyonian corals form soft, elastic colonies. Some receive the name "hand of death" thanks to the sinister look they can adopt.

Red coral
This is what red coral looks like when the polyps have their tentacles extended. For years this coral was collected to make jewelry and now it is quite rare.

Green velvety coral
Some corals form blocks of great dimensions, made up of thousands of small members of a colony. In the picture, each circle is a colony.

Stinging coral
When night falls, stinging coral spreads its tentacles to catch food, since they are armed with poisonous nematocysts.

27

MANGROVE SWAMPS

On coral islands, vegetable life usually begins with lichen, and then ferns and gramineae appear. However, sometimes the pioneer plant is a very curious tree—the mangrove.

Mangroves emerge from shallow water spreading their strange roots in all directions. In a short time, mangrove swamps are formed by strong trees that grow very quickly—about 2 feet (60 centimeters) in the first year.

The mangrove is well adapted to surrounding conditions: it is flexible and also able to withstand the high concentration of salt. It is one of the few trees capable of enduring the terrible tropical storms.

Mangroves have twisted and knotted roots that bury themselves in the clay and other air roots that bend upward and hang from the branches like lianas. The roots that go upward try to get oxygen from above the surface of the water.

They use different systems to reproduce. In most cases seeds are spread by the water, but some species of mangrove produce fruit that germinate on the tree itself and separate from the trunk, stick in the mud, and quickly give rise to another individual.

Mangrove swamps contribute to the formation of *atolls*, because their upward-bending roots make it difficult for water to circulate and accumulate sand.

Mangrove swamps are very fertile areas that provide a haven for much life.

(1) Roots above the water
Mangrove swamps are unique environments— the mangroves are the only trees able to live on soil covered in the sea.

(2) Reversed roots
To get atmospheric air, mangrove roots grow with their ends upward.

(3) A good refuge
Mangrove roots are a good refuge for many sea animals, such as these long, sharp-spiked sea urchins.

(4) The violinist crab
The violinist crab is a typical inhabitant of mangrove swamps. It uses its great claw to detect vibration just like strange hearing.

(5) The mud hopper
These large-eyed fish move on land by pushing themselves along with their pectoral fins. They can live out of water for up to two days.

THE OCEAN BEDS

On flat ocean beds, stones and detritus accumulate, as well as pebbles and coarse sand. The scenery becomes poorer deeper underwater because there is little light for most seaweed.

From 320 to 3,280 feet (100 to 1,000 meters) underwater, light disappears and vegetables can no longer live below 3,280 feet (1,000 meters), where there is total darkness. The temperature is very cold, since it falls progressively until a thousand meters underwater; from then on the temperature is constant.

Food is very scarce at greater depths. At 19,685 feet (6,000 meters) animal plankton is one thousand times scarcer than on the surface.

On the deepest beds, conditions change very little during the year, since there are no significant changes in temperature and storms hardly have any effect. Another of the most extreme environmental factors is pressure, as it increases by one atmosphere for every 33 feet (10 meters) down. This means that at 6,562 feet (2,000 meters) depth, the pressure is 200 times greater than on the surface. There are very few creatures capable of withstanding this kind of pressure.

Due to the difficult environmental conditions, creatures that live in the depths grow slowly, such as the bivalve Tindaria callistiformis, which reaches its maximum size (a few millimeters) after 100 years.

Underwater oases
On the ocean ridges, more than 6,562 feet (2,000 meters) deep, there are strange chimneys of boiling water that is full of underground products, particularly sulphur, which converts the immediate surroundings into oases of life. Here we find worms and giant crustacea, a large number of fish, and squid. The seaweed are substituted by peculiar bacteria that feed on the sulphur.

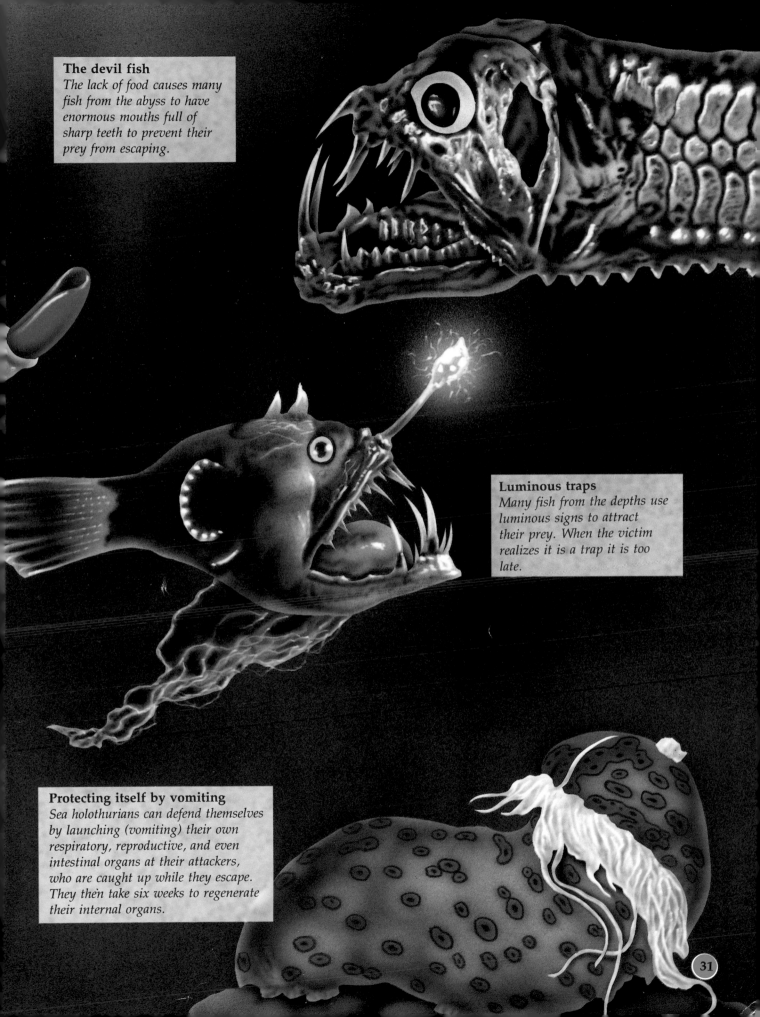

The devil fish
The lack of food causes many fish from the abyss to have enormous mouths full of sharp teeth to prevent their prey from escaping.

Luminous traps
Many fish from the depths use luminous signs to attract their prey. When the victim realizes it is a trap it is too late.

Protecting itself by vomiting
Sea holothurians can defend themselves by launching (vomiting) their own respiratory, reproductive, and even intestinal organs at their attackers, who are caught up while they escape. They then take six weeks to regenerate their internal organs.

Glossary

atoll a coral island consisting of a reef surrounded by a lagoon

chlorophyll the green coloring matter of plants

photosynthesis a process in which green plants synthesize organic material through carbon dioxide, using sunlight as energy

reef a chain of rocks or ridge of sand near the surface of the water

seaboard the shore or border of the land next to the sea

spawn to produce eggs in aquatic animals that will result in numerous young

stipe the stem of a plant that supports the cap of a fungus

symbiotic the living together of two different organisms in a mutually beneficial relationship

Index